T0146831

Extinction & Evolution

Jack R. Kryder

EXTINCTION & EVOLUTION

iUniverse books may be ordered through booksellers or by contacting:

iUniverse
1663 Liberty Drive
Bloomington, IN 47403
www.iuniverse.com
1-800-Authors (1-800-288-4677)

ISBN: 978-1-5320-0160-4 (sc)
ISBN: 978-1-5320-0161-1 (e)

Library of Congress Control Number: Pending

Print information available on the last page.

iUniverse rev. date: 07/14/2016

CONTENTS

Important Notes

Number Sizes

THROUGHOUT THIS BOOKLET, YOU'LL FIND the sizes of _billions_ and _trillions_ being used. Also the prefix _tera_ (a unit prefix in the metric system denoting multiplication by 10^{12}) and _peta_ (a decimal unit prefix in the metric system denoting multiplication by $10^{15)}$.

Please keep in mind these sizes are very, very, very large --- humongous really. Basically, they're sizes unfathomable to us humans but which must nonetheless be used to accurately state the sizes of the items mentioned. For instance, our current debt is one hundred twenty _trillion_ dollars. Since we have a very difficult time fathoming what a _trillion_ of anything is, we gloss over the size of the national debt --- much to our peril.

The Word Prodigious

The word/term _prodigious_ is used in many places in this booklet because the words fast, rapidly, big, large, Et cetera are just too inadequate. You'll definitely see why.

Prodigious—Associated Meanings:

extraordinary in size, amount, extent, degree, force · enormous · huge · colossal · immense · vast · great · massive · gigantic · mammoth · tremendous · excessive

PREFACE

THE INFORMATION IN THIS BOOKLET came from the United States government, learned scientists, concerned environmental groups, and many other knowledgeable sources. Based on my research, I've come to several alarming conclusions. I've explained the facts and conclusions in my own words to help summarize the information and hopefully make it more palatable. The details, sources, and extra materials can be found in the *Research Appendix*.

If you want to get a head start and a succinct understanding of what alarming processes are going on (and will soon be directly affecting you and every other person in the world), read the first section in the *Research Appendix* titled **Extended Notes.**

More information --

Website: worldnews.earth

E-mail/contact:
jack@worldnews.earth

**It's the mission and goal of WorldNews
to prevent wars and bloodshed.**

Every country in the world is running out
of freshwater where it's needed to survive.
Every country. It's a worldwide phenomenon.
There is no question the freshwater shortage
is occurring and getting worse at a rapid rate.
This fact has been proven, recorded and well
documented for many years throughout the
world.

Have doubts or disbelief? See _Research
Appendix_ at the end of this booklet or get
more proof on **worldnews.earth**. It's really a
twofold problem:

1. The earth is almost out of usable and
   available freshwater.
2. The freshwater that is available is very
   poorly distributed throughout the world's
   populations. The United States still has
   some freshwater; Saudi Arabia has none.

Most countries are somewhere in-between. Because of weather and hoarding, the situation concerning lack of freshwater locations changes daily.

People have been screaming and hollering for decades for somebody to _please_ fix the disappearing freshwater problem before it's too late. Very few have responded. I've got news for everybody—it's already too late. Most efforts represent a stop-gap at best. Again, check the facts and related information.

The lack of freshwater for survival brings up a very big problem. We are all programmed to survive. When our survival is threatened, we fight. Fighting leads to wars and bloodshed. Your blood, my blood. Again, It's the mission and goal of WorldNews to prevent wars and bloodshed.

If we can convince each other (all 7.3 billion of us) that what is happening is unavoidable and unstoppable, then maybe we can keep to ourselves and die with dignity when the time comes. No one likes to suffer but you really will have no choice.

Pretend you're shipwrecked on a small deserted island with no food, no water and no chance of rescue. You will have no option but to suffer and die—without affecting anyone else. When the time comes and the water disappears, we all have to act like we're on a small deserted island . . .

The earlier we start preparing ourselves, especially the younger ones (those under fifty?), the better chance we have of preventing wars and bloodshed.

I give a much broader picture of what I think is happening in the world today in this booklet. But the most pressing happening that overrides all other happenings is the disappearance of freshwater where it's needed for survival.

Disclaimer:

There are many websites referenced in this booklet. Before publication, each one was checked for availability and presence. However, the Internet being what it is, websites disappear from time to time. Obviously, I have no control over this. All of the listed websites were informative, interesting, and pertinent to this booklet. I hope none of them get removed from the Internet before you have a chance to refer to them.

INTRODUCTION

WHEN A PERSON IS FIRST diagnosed with a terminal illness, for most, their initial reaction is normally denial (and maybe some depression thrown in). As their symptoms progress and get worse, they are forced to a realization of their condition and say, "Uh-oh, I have a terminal illness and I better get busy and do what I can to stop it, if anything, or at least get my final papers in order."

Today our planet has somewhat of a terminal illness called _humans_. We've been gouging and eating into our home called Earth and consuming its resources and polluting it on a massive scale starting thousands of years ago. Our populations have spread throughout the earth like a cancer spreads throughout a human body. As far as the human race is concerned, the Earth is dying because of us. Because resources are fast disappearing as we consume them, we're just flat running out of an ability to maintain a sustainable and survivable life. Sad but True!

I've done my best in this booklet to give you enough information and facts, and without giving you too much verbiage, to convince you _extinction_ really is on the way and its severe symptoms and its very uncomfortable effects are just around the corner.

The symptoms are already affecting a large proportion of today's population in every country on earth. Can you think of one country --- just one country --- that is without poverty and denial of basic sustenance (food, adequate housing, proper sanitation, Et cetera to a certain percentage of its population? In some countries this denial of sustenance's involves and affects a large percentage of their population. The countries in Africa, the countries in South America, India and China come to mind --- _billions_ of people. Within the next several years, five at most, the rest of us, including you, will start feeling the symptoms of scarcity.

Because of our human nature, there is nothing we can do to stop it. We're going to continue to live as we are and to do everything we know how to do to keep surviving as long as we can.

And that's a good thing.

In addition, I know this is a very scary subject to confront. But just because you're afraid or scared doesn't make the fact of impending extinction go away.

Most people throughout the world who have been willing to acknowledge and deal with these problems and processes are universally attempting to stop them, reverse them, and make things better.

The Purpose of this Booklet

The purpose of this booklet is to reverse this thinking, to get you and all others around the world to stop attempting to fix the unstoppable and start finding ways today to avoid massive conflicts and wars that are sure to come. Let's use the knowledge of the impending extinction to our advantage.

Inaction and denial are the worst possible reactions you can have to the facts and conclusions in this booklet.

Evolution

Mentioning evolution in this booklet is an interesting sidelight. I'm in love love love with technology of any type or in any area it's being used in. I'm especially fond of computerized anything and what, how and where computerization is being used. I could have left the facts of our evolution to mechanical humans out of this booklet. But because so many millions of people around the world are being inundated and impacted with technology and loving it, I felt it appropriate to include the evolution to mechanical humans. I hope talking about evolution might make extinction seem less harsh and more believable. If you could care less about technology, by all means skip the parts about technology and concentrate on extinction --- the real focus of this booklet.

Please do not allow the talk of evolution to muddy the waters where stopping wars over extinction is concerned.

Jack R Kryder 6-21-2016

CHAPTER 1

Extinction Is on the Way

I'M GOING TO MAKE A prediction: I estimate that the human race as we know it today will be essentially extinct in fifty years or less. However, the extinction process is well underway today. In ten years or less, we (the world) will all be in big trouble. We're destined to be replaced by mechanical humans [mechanical humans are explained later in the booklet]. There is much evidence supporting the idea of impending extinction:

- The planet is running out of raw materials (minerals, oil, Et cetera) that we need to sustain life, materials that were deposited in the Earth millions of years ago and are irreplaceable. And it's happening at a prodigious rate.
- Another problem with minerals is where they're found/located. China has most of

the Earth's rare earth elements. The United States needs tons of these elements to manufacture many products, including electronics and batteries --- especially batteries for the new electric cars being developed. China is now starting to produce the same products. If China sucks up all the rare earth elements, where will we get ours? This same problem is cropping up with many other minerals as well.

- It gets even more complicated with minerals. For instance, using China as an example again, rather than buying minerals from producers, China buys the mines and the source of the minerals. Now they're in control and have ownership so they can ship the minerals to China. They're doing this worldwide. In the case of China, I think they do it with United States dollars. [Note: they own several *trillion* dollars of our debt, which we pay interest on. They use this interest money to buy companies and resources all over the world. Isn't that neat and clever of them?]

- The gist of it all is no one country has all the minerals they need to survive. As long as we all share, everything is A-OK. But when minerals get scarce, fighting occurs. Bad, bad, bad.

- The oceans are getting polluted and acidified at a prodigious rate and are almost completely devoid of edible fish, which are not being replenished. Many countries, such as Japan, rely on marine life as their only source of protein. Fish gone, protein gone.
- Freshwater that took millions of years to accumulate is being consumed at a much faster rate than it's being replenished. Pollution of freshwater is also happening at a prodigious rate. This problem is worldwide. No freshwater, no life.
- Food production is waning at a prodigious rate. Because of human development, arable land is disappearing at a prodigious rate. Because arable land has been farmed for so many years (in some areas for thousands of years), it is bereft of necessary nutrients to grow crops, and farmers must use fertilizers. If it weren't for the fertilizers, along with pumping up all our groundwater and depleting water tables to water crops, most arable land would be a barren wasteland. This is going on worldwide. We're close to nobody getting any food. In addition to water running out, fertilizers are a special concern.
  - o Fertilizers are complex and power-intensive to manufacture.

- o To get them where they need to be and apply them uses a lot of power, which in turn pollutes the air.
- o Their manufacturing processes help pollute the ground.
- o Agricultural use causes runoff that pollutes the surrounding soil and water.
- o They are unavailable in many parts of the world. As a substitute, many countries resort to the age-old solution of using animal or human feces.
- o They only supply about half the types of nutrients that were originally in the soil. Therefore, we are not getting balanced nutrients in the food we eat that is grown with fertilizers.
- o MOST IMPORTANT. Their production uses up many of the world's resources at a prodigious rate.

- Climate change is wreaking havoc all over the world. Too much water where we don't need it. No water where we do need it. Weather all out of whack worldwide with higher temperatures in both winter and summer in some areas, and in others, colder temperatures than normal. Huge wildfires. Tornadoes in the United States in large numbers where they've ever been before. Ice melting all over the planet, which is diluting oceans and causing flooding in low-lying areas of the world. The list goes on and on --- right?

- There is also evidence the climate changes and other natural phenomena that are wreaking havoc all over the world are more due to the natural cycling of the Earth than to human activities. A couple of examples are active winds: heavy winds on the ground, including tornadoes, and upper winds like the jet stream changing speeds and positions. El Niño and La Niña in the oceans affect weather on land. Et cetera.

- We're making our planet uninhabitable for humans. Even if humans stopped polluting the atmosphere, freshwater, oceans, and soil 100 percent, it would make no difference in what's happening because the damage to these areas has already been done and can't be reversed. Horribly, the damage continues to get worse.

- Today, as opposed to the future, the world is way overpopulated as far as our ability to support everybody (with jobs, food, water, sanitation, housing, Et cetera). And it's getting worse day by day. Check out countries like the United States (yes, the United States), India, and China. Or check out the continent of Africa. These areas have never even come close to meeting their general populations' needs except the United States. And even we fail miserably.

- Our infrastructure (electrical grid, satellites, cell towers, Et cetera) is extremely vulnerable to solar flares and coronal mass ejections (CMEs) that are predicted to happen in the near future. No electricity in the United States, no living our lives.

- Many of the satellites we rely upon for civilian, scientific, and national defense use are nearing the end of their operating life spans, and plans for replacements are scarce. No weather reports, no GPS, no meeting military needs, no analyzing crops or water conditions, or many of the other things satellites are used for.

Demands for Resources

If we assume that all people will live at a particular standard of living, there is a finite carrying capacity of the Earth, above which population growth will not be sustainable because of rapid depletion of resources and too much pollution. For example, is it possible for all those currently alive to live at what is called a "Western middle-class" standard? --- No. To do so, we would need more than four Earths to supply the resources and assimilate pollutants.

Here's an interesting extrapolation. The United States has about 300 million citizens. China has about 1,200 million and so does India. Both these countries are waking up and declaring they want the same level of life people in places like Japan, Europe, and the United States have. And they're gearing up to make it happen.

Presently, the United States uses about 25 percent of worldwide produced resources per year, and the rest of the world uses the other 75 percent. When China starts using resources at the same rate as the United States, they will be demanding 100 percent of the world's resources per year.

In other words, because China has four times the population of the United States (300 million vs. 1,200 million), they will need four times the resources: 4 x 25% = 100%

Now factor in India. They will also demand 100 percent of the produced resources. Something is going to have to give from all countries involved. Who will go first --- the United States, China, India, countries in Europe?

Will any country give up their way of life, or their projected way of life, without a fight and going to war?

And think of how fast resources will be used up per year. Depletion of resources will get prodigiously accelerated. Will any resource we need to survive last very long into the future? The answer is no! What do you think will happen when there are not enough resources to go around? I'll tell you --- wars, wars, and more wars. Bloodshed by the thousands of gallons.

<u>That is, unless we do something, starting today, to stop the wars and bloodshed.</u>

Partial List

The above litany of coming problems is just a partial list of the problems humans will be facing in the near future.

I think of all the above problems, freshwater is the most critical --- no freshwater, no life. Period. At the very least, there'll be no drinking water, no food production, no sanitation, and very few manufacturing processes.

The only thing I'm aware of that might be used to create freshwater is large desalination systems. It will require thousands of them to keep us in freshwater. We have neither the time, the material resources to build them, nor the power needed to operate them (electric generators, coal, natural gas, atomic reactors, uranium, or whatever). Also, they need to be near an ocean to get the saltwater they need to desalinate. Therefore, they will be far away from most populated areas that need the freshwater, and it will be almost impossible to get the water to people where it's needed.

It may be somewhat easier to locate the desalination plants where they're needed and pipe water from the nearest ocean to them. But even this scenario poses problems. And it still doesn't obviate the need for large amounts of power to pump the water and power the desalination plants.

There is freshwater trapped in ice in many places throughout the world. Most of it is melting and going straight into the oceans. It would be almost impossible to capture ice water and transport it to where it's needed. Even if that process succeeded, it would only be a stopgap measure because there is very little new ice being replenished, thanks to global warming. In fact, it's global warming that's causing the melting.

The scary thing about freshwater disappearing is that it's a worldwide phenomenon.

There are several very sad, unequivocally undeniable facts about all of the above-listed problems:

- As I said earlier, the above is only a *partial* list of problems we face.
- All of the above can be verified by using appropriate books, periodicals, television, newspapers, and of course, the wonderful Internet (reference the *Research Appendix*).
- All that's happing is ongoing and irreversible.
- Our human nature is a huge predictor of our coming extinction. We're unwilling to sacrifice and unwilling to give up our hard-won way of life. A good example of this is the United States debt. We all know the United States is bankrupt, yet we continue to borrow at a prodigious rate to sustain our way of life (we're up to borrowing several *trillion* dollars a year, and the current debt is twenty *trillion* dollars in bonds and close to one hundred *trillion* dollars in entitlements!). And since we're becoming a socialistic country, attempting to meet everyone's needs (in some cases even outside the United States), the borrowing will only increase and continue to the point where it finally crushes us and

we have to default. And watch out, this is about to happen in the near future.

- Besides the United States government having debt problems, corporate America is borrowing money faster than they're making it. Another formula for disaster.
- Again, our human nature is a huge predictor of our coming extinction. The Earth's irreplaceable resources needed to support and sustain human life are fast disappearing at a prodigious rate, yet we continue to gobble them up as if their quantities were unlimited. Our attitude is let others sacrifice while I continue with my lifestyle and way of life.
- Human nature! We basically have our heads buried in the sand, ignoring what's happening around us and around the world that is in the process of making the planet uninhabitable for humans. Sad, but oh so true --- don't _you_ think this is sad?

CHAPTER 2

Free Money

I'VE HEARD A LOT OF people say that fixing many of the above problems will take a lot of money, money we don't have. **Wrong!** Money is free and always has been. Money is basically a _concept in our minds_. Money is no more than:

"What am I willing to accept for me to give you what I have or what you want?"

Money has taken many physical forms over the years, including pebbles, gold, various other metals, chickens, cows, property, paper currency, an entry in a computer, Et cetera. If a government wanted to, they could give their citizens money by printing paper currency and passing it out. As long as the currency is recognized and accepted as money, then everybody in that country would have the money they needed to live on, buy a house, buy food, buy clothes, and pay bills. This is only one form of printing money. Economists call the process of creating money out of thin air _printing money_ even though much of the free money created today involves computers rather than currency.

Example: This was done by Germany in the 1920s and 1930s with hyperinflation as a result. In today's world, we can generate all the money we want in a fraction of a second by entering the amounts in a computer and honoring those entries. Simple as that. [Author's note: In the United States we print money all the time, mostly through our banking system and the Federal Reserve System.

Why don't we? It would cause _inflation_, chaos, and catastrophe. If everybody all of a sudden had money, where would all the physical resources and people come from to produce all the food required or get materials to build houses or make enough clothes or manufacture automobiles? Or for that matter, if we create an _infinite_ amount money, where would we get all the physical resources, people, and time needed to fix all the world's fixable problems?

The world's economies have always been structured like a pyramid to support humankind. The richest at the top and others are poorer as you approach the base --- with most of us somewhere in the middle. Why? Because it has always taken all kinds of talents, capabilities, menial tasks, trades, and professions to support humankind. Just think about here in the United States. How many thousands of types of jobs/activities need to get done each second of each day for all of us to survive? If everyone was at the top of the pyramid, who would do all the necessary jobs so we could survive? Everything would grind to a halt.

One interesting fact about creating money. Banks around the world do create free money every day and distribute the money through loans. It adds *billions* of dollars to existing economies on a daily basis by simply making entries in a computer.

Money creation

From Wikipedia, the free encyclopedia

https://en.wikipedia.org/wiki/Money_creation

> **Money creation** (also known as credit creation) is the process by which the money supply of a country or a monetary region (such as the Eurozone) is increased. Typically, central banks create money by using notes and manufacturing coins. However, in the contemporary monetary system, most money in circulation exists not as cash or coins but in an electronic

form, which is created by commercial banks.

Commercial Banks create money in the form of <u>demand deposits,</u> when they make loans to households or companies. When a bank makes a loan, a deposit is created at the same time in the borrower's bank account. In that way, new money is created as a bookkeeping entry, with the loan representing an <u>asset</u> and the deposit a <u>liability</u> on the bank's balance sheet.

Ultimately, the amount of new checkbook money created through lending is a large multiple of the initial deposit. The key word here is deposit, meaning available free money. And it's all very legal, effective, and useful in the support of economies --- to the tune of <u>billions</u> of dollars.

The United States Federal Reserve controls the amount and velocity of money circulating. At any given time, they can increase it (that is create money) or decrease it (eliminate money). They use this control all the time.

Around the world, any amount of money can be created at any time to meet emergencies. Wait and see. You and everyone else will consider extinction an emergency.

What I'm saying is because money is free and we can freely generate as much as we want, then money is out of the _equation_ in our hopeless effort to survive. Infinite money is unable to buy what is nonexistent. For instance, when all the freshwater is gone, it's gone.

CHAPTER 3

Prognostication

ALL OF MY POSTULATING AND predictions may sound like I'm being a harbinger of doom. ***Nothing could be further from the truth***. I'm being much more of a _prognosticator_ and predicting our extinction as part of our evolution from flesh and blood humans to mechanical humans. For instance, is our extinction good or bad? I say neither.

Extinction is nothing more nor nothing less than a normal progression and part of Earthly evolution taking place.

The case for evolution --- from humans to robots to humanoid robots to mechanical humans.

By now, it's an accepted fact that all species on Earth evolved. Nothing was created instantaneously or spontaneously. It has all been based on evolution. There are pretty obvious differences between plants and animals, but --- at the chemical level --- the cells of all plants and all animals contain DNA in the same shape, the famous "double helix" that looks like a twisted ladder. What's more, all DNA molecules --- in both plants and animals --- are made from the same four chemical building blocks called nucleotides. We all require water, Et cetera.

The resources needed to support DNA-based species are fast disappearing and that is leading to our extinction. Obviously, if any evolution of intelligence is to continue, it will have to be based on something new and different. The only thing I see coming along to replace DNA-based species are intelligent computer-based electromechanical systems.

Let's say for a moment you buy into the fact that DNA-based species are rapidly becoming extinct, and that evolution wants to pass on the current evolved capabilities of intelligence to something that *will* survive and continue to evolve. It's only within the last several hundred years, when the Industrial Revolution really got going, that this process of evolution from humans to machines has become visible and evident. Look around you. We've gone from dumb machines to smart machines. Check out how many smart machines/devices you have around you and how powerful and smart they are. The timing is too perfect to ignore. I believe the fact that humans are becoming extinct, and that we have created our successors just in time to pass on intelligence, has nothing to do with happenstance.

The evolution that is happening today started happening on day one.

CHAPTER 4

My Life Regarding Technology as It Relates to Evolution

I STARTED MY LIFE IN computers in 1960. I've followed the development of computer systems and related subjects fairly well over the past fifty-six years. I've seen their capabilities expand and develop at an exponential rate. I saw electronics go from vacuum tubes to a single transistor to small pieces of silicon with _billions_ of transistors on them.

Intel is a company that has been developing central processing units (CPUs) for years. In 1971, the first generation Intel processor sported **2,300 transistors** and ran at a paltry 740 kilohertz (kilo = thousand, hertz = cycles per second).

NVidia, in 2016, produces the Tesla P100. At **150 billion transistors,** it is one of the largest chips ever made and may be one of the fastest. It runs at 21.2 *teraflops*. A *teraflop* is a measure of a computer's speed and can be expressed as:

- A *trillion* floating-point operations per second.
- That is ten to the twelfth power floating-point operations per second.

In the near future, computers are expected to operate at *petaflop* speeds. A *petaflop* is a unit of computing speed equal to one thousand million (10^{15}) floating-point operations per second.

And to think, all this has happened in just the last fifty-six years. We're definitely on a roll!

The brain of Homo sapiens has an estimated hundred *billion* neurons and runs at a very, very paltry 1 kilohertz (one!) --- which is very, very, very slow. Tesla P100 systems do thousands of simultaneous processes at speeds *billions* of times faster than our brains. In addition, these CPU systems are being taught/programmed to be many thousands of times smarter and resourceful than the human brain, and the development is happening at a prodigious rate. Some people refer to this as artificial intelligence, but the fact is that intelligence is intelligence. Just because the intelligence resides outside our brains is irrelevant. The capability and usefulness of these new systems are getting beyond our ability to comprehend and wrap our minds around them.

Solid state memory units now go as high as one *trillion* bytes. I saw magnetic disc storage units go from one million bytes to several *trillion* bytes.

The amazing thing is these CPU systems and storage units range in size from one-quarter inch square to about five inches square. There's a tremendous amount of computing and storage power crammed into very small spaces. In most cases, much smaller in size and weight than our brains but many, many, many times more powerful.

There is much precedence for humans developing equipment with capabilities that we need to survive that had to go far beyond our puny selves:

- When we needed to grow large amounts of food, we went from horse-drawn plows to giant tractors pulling large wide plows.
- When we needed to transport ourselves and materials, we developed automobiles, trucks, trains, large ships, large earth movers, airplanes, Et cetera.
- To build large buildings, we developed large cranes that are capable of lifting tons of materials high in the air.

- When we needed to see details our naked eyes are not capable of seeing, we developed optical microscopes, electron microscopes, binoculars, telescopes, Et cetera.
- We have developed extremely powerful and efficient weapons for killing and destroying our fellow man --- weapons much better than knives, swords, and axes.
- We've created very sophisticated medical equipment.
- We created manufacturing machines that produce millions of products a day in all areas—pills, food, clothes, cars, cell phones, integrated circuits, wood, paper, plastics—on and on.
- When we wanted to communicate farther than to our next-door neighbor, we developed sophisticated communication systems and capabilities.

You get the idea. So why should anybody be surprised we're developing very advanced powerful computer systems, other electronic products, and mechanical humans that help us get things done that we're unable to accomplish as puny, vulnerable, limited human beings?

And realize most of this increase in CPU systems and ubiquitous super-powerful electronic capabilities may have started many years ago, but the majority has taken place in the last fifteen years. Things have been proceeding at an accelerated rate. One of the reasons this is happening is because in 1960, when I started, there were maybe 25,000 people working in these fields. Now we have millions all around the world working in these fields --- including young kids. It's a process where it seems faster and better computers and more people helps us design and produce faster and better computers --- and many other products as well --- from bridges to aircraft to medical equipment to drones to clothes to cell phones to computers to . . . you name it. And of course, computers are helping us do it all.

For examples of companies developing and advancing these technologies, check out the capabilities, functions, and future products of companies like Amazon, Google, Facebook, Intel, Microsoft, Apple, and Cisco Systems. There are thousands more companies in all fields coming up with amazing technologically advanced developments as well.

I would be remiss if I didn't mention the defense industry around the world that down through the ages have contributed mightily to technology and products we use in our lives today (even though they originally developed the technology to kill people or to keep us from being killed, we managed to turn things around and use the technology for good purposes).

If you really want to check something out that is getting bigger, faster, more powerful, more complex, more useful, and more ubiquitous, look at the millions of interconnections on the Internet and the hardware it takes to accomplish this. As they say, the beat goes on.

CHAPTER 5

Current Advances

TO ILLUSTRATE, HERE ARE A few of advances that are coming along right now:

<u>Super Flat Material Could</u> <u>Extend Life of Moore's Law</u>

Moore's law states that every eighteen to twenty-four months, processing power will double, and it has been steadily observed to be true since 1965, enabling the rapid technological progress over the last four decades. Researchers could be fending off the demise of Moore's Law with the help of a new kind of flat semiconducting material made of tin monoxide that is only one-atom thick, allowing electrical charges to pass through it _faster_ than silicon or other 3D materials. One atom thick --- can't get any thinner than that --- wow.

Another Example of Extending the Life of Moore's Law

Fears that Moore's law --- which dictates the exponential growth of processing power --- would falter this year have been allayed after IBM revealed processors with circuits just 7nm wide (nm means nanometer --- *one billionth* of a meter --- very, very small indeed).

Chance Discovery Puts Graphene Electronics Closer to Mass Production

We've heard plenty on the wonderful properties of graphene, but the super-material *par excellence* still hasn't found its way to commercial products because (up to now) it is too delicate for real-world conditions. Now, in a lucky and perhaps game-changing discovery, scientists at the Brookhaven National Laboratory (BNL) have found that placing graphene on top of common industrial-grade glass is a cheap and effective way of making it resilient and tunable, paving the way for the production of graphene-based electronics on a large scale. Possible applications for graphene-based electronics include better solar cells, OLEDs, batteries, and supercapacitors, as well as *faster microchips (including computers)* that run on very little power.

Researchers Use Coal to Make Electronics

Using the molecular complexity of this highly varied material, we will soon be harnessing the real value of coal's diversity and complex chemistry. Coal could become the basis for solar panels, batteries, or electronic devices.

Will Humanoid Robots Build Tomorrow's Aircraft?

Humanoid robots with human capabilities, including specially designed hands and feet with complete mobility and dexterity, are being designed to build aircraft. They do many jobs quicker and more safely than humans. They also go way beyond the capabilities of robots that help build automobiles.

Mapping the Thesaurus of the Human Brain

Researchers have used MRI data to create a map of how the brain organizes language.

Mass Storage --- Very Small Glass Disc --- That Lasts Forever

Newly developed glass discs are reportedly capable of storing up to 360 terabytes per disc of data and will last almost forever at room temperature (or as long as the current estimated 13.8 *billion* years our universe will keep going on). Three or four disks will give you more storage than you'll ever need, and it lasts forever --- a very, very long time indeed.

MIT

Some articles in the March/April 2016 issue of the Massachusetts Institute of Technology's *Technology Review*:

- Google's dream machine. Researchers home in on a workable quantum computer. [Aside: In physics, a **quantum** is the minimum amount of any physical entity involved in an interaction. For computer development, read tiny, tiny, tiny, but very powerful and fast and uses almost no power.]
- Robots Teaching Robots. Breakthrough. Robots that learn tasks and send that knowledge to the cloud for other robots to pick up later.

- Tesla Autopilot. The electric vehicle maker sent its cars a software update that suddenly made autonomous driving a reality.

Advertisement

An advertisement by a large technology company named Qualcomm:

We started by connecting the phone to the Internet, now we're connecting the Internet to everything. By inventing technologies that connect your car, your home, and the cities in which we all live, we're accelerating a smarter, more seamless and intuitively synchronized world.

CHAPTER 6

Computer Systems Are Evolving into Mechanical Humans That Will Eventually Replace Humans

I THINK WE CAN ALL agree computer systems and related artificial intelligence are very powerful today. Computerized intelligent systems are ubiquitous:

- We talk to our cell phones and computers; they understand us and do our bidding.
- Cars drive themselves and do a better job than humans. Almost all accidents involving self-driving cars and humans happen because the human drivers were violating the rules of the road, while the self-driving cars were strictly following them. And in some cases, the humans were violating common sense and safety

as well. Get humans out of the equation and things get better.

- Commercial aircraft take off, get us from point A to point B, and land us at point B, all with no action or intervention needed from the pilot. And they do it more safely. Get humans out of the equation and things get better.
- Small computer systems control many of the functions within our cars, homes, and businesses.
- Millions of robots around the world are performing jobs better than humans, and in many cases, doing jobs humans are incapable of performing. Get humans out of the equation and things get better.
- Hundreds of thousands of smart computerized instruments have been designed for use in all areas that support human life. A good example is the medical industry.

There is much evidence that these capabilities are taking control of the world. And it's all happening today as opposed to some distant future date. So that's some of the things going on now. Here's a bit of what I see coming to fruition as far as computing systems are concerned in the near future:

- In the next ten years or so, computer systems will far surpass our form of intelligence and consciousness and be many times more powerful and far-reaching than ours.
- So-called artificial intelligence is being developed at a very prodigious rate (as I said before, no intelligence is artificial --- intelligence is intelligence).
- Electronic memory, for all practical purposes, will be free and basically infinite (it's almost that way today). All of our human knowledge will be available to them (we're in the process of digitizing everything).
- Their circuits are already millions of times faster than our puny brain synapses, and they're getting faster every day. The number of memory units available to them, collectively, will exceed the number of our brains' memory cells by many millions of units.
- In addition to having almost infinite memory units, their memory units never forget, and they have almost instant recall for any information they need.

Google says, "We're proud to have the most comprehensive index of any search engine, and our goal always has been to index all the world's data." I just did a search on Google using the word *computer*. The result was:

1,830,000,000 (1.8 *billion*) results in 0.55 seconds (basically instantaneous).

Pretty fast and comprehensive, huh? Hell, if computer systems can do that today . . . well, there's literally no telling how much computer systems will be able to do in ten years --- let alone tomorrow.

I think technology in general is now at a critical mass. We're creating so many new smart products and so many new intelligent capabilities in hundreds of areas so fast that none of us can keep pace. Their smarts, capabilities, and intelligence are overwhelming us.

Their mechanical bodies will take many useful forms that far surpass our human form (think reproduction on a large, varied scale --- varied is the key word here --- and with no labor pains involved). The type of sensors available to them will make them aware of many more things around them than flesh and blood humans are natively capable of sensing.

And maybe the most significant capability, they will all be interconnected using technology like cell phones, Wi-Fi, and the Internet all intrinsically built in --- no external devices needed. So collectively, they'll all know everything and help and control each other to get things done. No secrets or protected information needed or expected.

Basically, what we're trending toward is computerizing everything and, most impressively, creating advanced **mechanical humans**. Notice the term mechanical human is _waaaaay_ better and more accurate than the term robot. We're making the transition from flesh and blood humans to robots to humanoid robots to mechanical humans. The term robot has the connotation of a dumb, single-minded intelligence and basically has a single simple repetitive function to perform. Humanoid robots do better. But mechanical humans will go far beyond being mere robots and especially way beyond ourselves.

CHAPTER 7

Evolution in Motion

ALSO, NOTICE THERE IS A continuity here. Just like us, mechanical humans will be able to reproduce and have fantastic intelligence. We'll be passing on the very essence of the flesh and blood human species, including these _unique_ qualities:

- Super intelligence compared to any other creature on Earth.
- A very general mental capability that among other things involves the ability to reason, plan, solve problems, think abstractly, comprehend complex ideas, learn quickly, and learn from experience.
- The ability to learn from, understand, or deal with new or trying situations.
- The ability to apply knowledge to manipulate one's environment or to think abstractly as measured by objective criteria.

- Mental acuteness.
- Capacity for learning, reasoning, understanding, and similar forms of mental activity.
- Aptitude in grasping truths, relationships, facts, meanings, Et cetera Manifestation of a high mental capacity.
- Consciousness and awareness of who we are and what we are.

Flesh and blood humans are very high-maintenance as far as existing on Earth is concerned. Mechanical humans will require much less in the way of resources to flourish on Earth than flesh and blood humans. No need for flesh and blood human food, no water, no oxygen, no bodily wastes, Et cetera The really key point to observe is that mechanical humans will need *different* materials and processes to survive than we humans need --- materials that aren't starting to disappear here on Earth. They won't be nearly as subject to environmental concerns such as heat, cold, air quality, and time constraints. As you might now be aware of, the list of *don't needs* of mechanical humans is very long. Who knows or can predict how much more mechanical humans will develop and thrive on Earth after us flesh and blood humans become extinct --- or even before then?

CHAPTER 8

Relevant Facts about Extinction & Evolution

TO ME, THIS ALL SEEMS like an evolution that was meant to be. As I said before, the evolution from flesh and blood humans to mechanical humans was set in motion from the very beginning of time. Here are the main relevant facts as I discern them:

- This evolution is neither good nor bad --- it just is.
- Human extinction and the evolution to mechanical humans has been happening simultaneously from the very beginning. It was part of the original design.
- We're now at a critical point where it is too late to stop extinction, and it is too late for us to stop development of mechanical humans. We're having too much fun

developing all this new technology, we're addicted to technology, and we've become like a freight train barreling down the tracks. Anyway, it was preordained.

- We'll develop mechanical humans to the point where they're smart enough to reproduce and get along without us, and that will be around the time we start disappearing.
- Evolution is inevitable and unstoppable.
- We need to stop deluding ourselves and face reality.
- We need to accept extinction and get used to it and stop attempting to fix things --- *e.g., let's stop attempting to prevent climate change, loss of freshwater, depleting the oceans, Et cetera.* These types of activities are truly futile and a waste of time and energy. (Not a waste of money. Remember, money is free.)
- Stop worrying about future generations, and for sure stop having children. As I said before, we aren't even able to take care of the people already here. Why create more impoverished, disgruntled, and unhappy people? All procreating does is speed up the process of extinction.
- Rather than spending our time and efforts fighting the inevitable, we should be spending the time we have left making

those who are alive today (all over the world) more comfortable (including ourselves).

- Maybe most important, we need to devise ways to help future generations (mainly our children, grandchildren, and great-grandchildren --- i.e., those who are already alive) deal with the problems that lie ahead for them.
- When we know what's coming and prepare for it, we're much better able to handle it than when it's just dumped in our laps.
- A final thought. Let's stop and smell the roses and have a little fun along the way while we still have a chance.

A Final Plea

Many countries are in financial trouble or are heading for financial trouble (most notably the United States). However, money problems, if handled properly, can be solved using free money.

Extinction is something altogether different. Extinction, as part of evolution, is on the way. There is nothing we can do about it. What we do need to concern ourselves with are the end days that will occur between now and complete extinction. It involves the disappearance of food and water.

Basically, we can live with the fact we are all going to die --- nobody lives forever. If we lived for 800 or 1,000 years, we might get upset at losing 600 or so years of life. But knowing we only live for eighty or so years, we can live with losing a few years of life.

Starving is something else altogether. Right now, this very minute, several *billion* people around the world are without adequate food and water (including many in the United States). As time passes, during the next ten years or so, the problem will get severe and will spread to all of us.

We humans have an intrinsic inborn need to survive. When food and water start to become scarce, our natural instinct is to fight our neighbors to get the food and water we need to survive.

In Venezuela, things have gotten so bad that there is much infighting among citizens attempting to acquire just the basics to survive. Almost the entire continent of Africa is in this type of situation. The list goes on.

Between countries, things can also get bad. Syria is a typical example of what's going on as far as getting resources from other countries. In the June 13–June 26, 2016, issue of *Bloomberg Businessweek*, an article titled "When the State Wilts Away," had this to say:

Even Islamic State's political power may soon be affected by drought. As water levels in Lake Assad in Syria plummet, Raqqa, the group's stronghold, is facing severe shortages. Last year, Islamic State's press officer, Abu Mosa, told Vice News that it would consider attacking Turkey to gain access to additional water resources.

This type of survival activity has been going on basically since the beginning of time; one tribe or country attacking another to gain access to resources. History is a good example. In 1939, Japan was in dire straits as far as survival was concerned. They got so desperate, they attacked China to gain resources for survival. To prevent the United States from interfering with their attempted theft of resources, they bombed Pearl Harbor in Hawaii to destroy our naval power. We all know how that ended --- World War II ended up killing millions of soldiers and millions of innocent civilians. If you think World War II was bad, wait until you get involved in World War III. Nobody and nothing will be safe and able to hide from the carnage.

What would you do if you caught your neighbor stealing your car? Or came home to find a robber stealing all the food from your refrigerator? It matters not if it's neighbor stealing from neighbor or a county stealing from a country. A fight is sure to ensue. If we do nothing, there *will be* much bloodshed throughout the world.

I say if we start to plan now, we can set things up so we all share equally. That way, we'll avoid much fighting and bloodshed. This may sound like a utopia I'm describing. And it probably is. But if we can get the 7.3 *billion* people from around the world working on it, I feel it can be achieved. What is it they say? "7.3 *billion* people can't be wrong." Maybe I'm being naïve, but I think it's well worth attempting.

A Final Note

By the time things start to get critical, mechanical humans will be far enough along in their development to be a tremendous help in figuring out how we can share equally.

A Final, Final Note

Where survival is concerned, there is usually no middle ground. We'll either end up with a utopia or a dystopia. I vote for utopia.

CHAPTER 9

Condensed Bio: Jack R. Kryder

- I was born in 1940 and raised in San Pedro, California.
- My dad worked in the defense industry (shipyards and aircraft manufacturing).
- My mom worked as a clerk for various government agencies.
- I seemed to be born with an engineering/ science-oriented mind.
- I became interested in science when I was a kid by reading comic books, science fiction, biographies of famous scientists/ inventors, and watching *Flash Gordon* on television.
- In 1957, I started a formal interest in electronics by becoming a Radio Amateur. My call sign was K6PUX. My handle, assigned to me by my fellow Radio Amateurs, was _Puxy_ --- I hated that handle.

- In 1960, I joined the army. They sent me to boot camp, electronics school, computer maintenance school, and then over to Germany to maintain computerized missile defense systems.
- In 1963, I was discharged from the army and started electronic engineering school at Arizona State University.
- In 1966, I had to drop out of school for personal reasons. I managed to get a good grounding in mathematics, beginning electronic theory, and girl-chasing (had my first child in September 1965).
- Luckily for me, the electronics industry, including the development and applications of computers, was going gangbusters. Because of my army training and my ASU schooling, I had my choice of jobs.
- I started out as an electronic technician and soon, through on-the-job training, worked my way up to engineer status.
- I advanced from technician to electronic systems designer to computer designer and computer programmer to business software programmer and information technologist.
- Even though I remained a grunt for most of my career, I did get to work on systems and areas that contributed to the advancement of mankind (mostly in small ways by today's

standards). Besides, as I said before, someone has to do the grunt work.

- During all these activities, I helped raise two great sons.
- I retired in 2007.
- Since retirement, I've kept myself busy traveling, reading, watching a wide range of TV shows, using my computer for many tasks, searching out new restaurants, and having fun with my wife, animals, relatives, and friends.
- However, no Facebook, Tweets, texting, cell phone use, newspapers, TV news programs, or politics of any kind. That's just me. I have a lot in common with tortoises. We can pull in our heads, wrap our arms in front of the opening, and let the world go by. Heavy sigh.

All my life, I've been lucky enough to be rather eclectic in my interests. However, my two key interests have always been electronics (if it runs on AC or is run by a battery, I'm interested) and the health of our good planet Earth.

Inside Creates Outside?

When I was younger (in my twenties), I remember seeing pictures of Earth taken by astronauts traveling to the moon and thinking we designed our outer world to be like our anatomy:

- We have veins and arteries that have little red and white pods that carry food and oxygen where needed and pick up wastes along the way for disposal. Have you ever looked down on Earth from above? You'll see thousands of miles of *arteries* with big and little engine-driven pods scurrying around doing the same types of tasks as our red and white pods do (along with thousands of other tasks).
- We have nerve systems that connect our body parts to our thinking/control unit so all our body parts are connected and controlled properly. These nerves also tell us if things are working properly, and if not, they tell us to get help. Again, look down from outer space. Do you see all the metal wires, fiber optic cables, microwave radio antennas, Internet backbone, telephone connections, underwater cables, satellites, cell phone towers, TV cable systems, Et cetera? Are they not our external nervous system connecting us all together?
- We have city, county, state, and national governments. Don't they collectively

represent and function as our thinking/control unit?

- Body cells have walls for protection and a closed environment for them to get their jobs done. We have houses and buildings with walls for environmental control and protection so we can get our living and jobs done.
- We have different types of organs that perform various functions. Aren't cities and divisions of cities somewhat like organs?
- Our skin protects us, holds body parts in, and keeps us from looking too weird and ugly. Don't our clothes do that for us also?

OK, OK. Enough of the anatomy analogy. But it was the continuation of this analogy that led me to the conclusion we were continuing to clone our inner selves into the outside world when I put the following two facts together:

1. The fact we are programmed/designed to reproduce to keep the human species going.
2. The fact the way we are doing it is cloning ourselves in the form of mechanical humans.

It then seemed only natural to link the two facts up with evolution. There's been an evolution starting with amoebas that has continued forward to creating mechanical humans. As they say --- it's survival of the fittest. And mechanical humans are definitely going to be more fit to survive than we are for the coming times.

CHAPTER 10

I'm Known as *Jack the Happy Atheist*

WHILE I'M AT IT, I should mention one last thing: I'm an atheist and firmly believe there is no god. So I have no idea who or what created this universe, but my engineering/scientifically oriented mind does make me inclined to believe it was created.

All I know is that where we came from or why we were created is unimportant and unknowable. What has been important is we are interested in what exists, how it works, and how we can use this knowledge to develop capabilities to make ourselves comfortable and happy. And we've done a really good job of that down through the ages. It's how and why we got to where we are today. We were created and designed to work this way and achieve the things we have.

An Apocalypse?

Many people think we're in the middle of an apocalypse. Here's what Wikipedia has to say about apocalypse:

Apocalypse

From Wikipedia, the free encyclopedia

https://en.wikipedia.org/wiki/Apocalypse

*An **apocalypse** (Ancient Greek: ἀποκάλυψις apokálypsis, from ἀπό and καλύπτω meaning "uncovering"), translated literally from Greek, is a disclosure of knowledge, i.e., a lifting of the veil or revelation. In religious contexts it is usually a disclosure of something hidden.*

In the Book of Revelation (Greek: Ἀποκάλυψις Ἰωάννου, Apokalypsis Ioannou – literally, John's Revelation), the last book of the New Testament, the revelation which John receives is that of the ultimate victory of good over evil and the end of the present age, and that is the primary meaning of the term, one that dates to 1175.[1]

Today, it's commonly used in reference to any prophetic revelation or so-called end time scenario, or to the *end of the world* in general.

ON TV

In this recent TV program:

The Story of God
National Geographic
Morgan Freeman
S1 E2
4-10-2016
Episode's title: "Apocalypse"

Religions consulted:

- Judaism
- Christianity
- Islam
- Buddhism
- Hinduism
- Maya

Apocalypse program's conclusions [I'm paraphrasing]:

- Apocalypse means we'll be free of injustice.
- Apocalypse leads to a better world.
- Apocalypse has nothing to do with war.
- Apocalypse has nothing to do with death.

- Apocalypse has nothing to do with some far-off day of judgment.
- Apocalypse means to have a concentration on enlightenment.
- Apocalypse is a state of mind that helps us see the truth.
- These ideas about an apocalypse are here now --- no waiting involved.

I go a little further in that I believe we're in the midst of an apocalypse in the sense there is going to be an end of humanity as we know it today and *that of the ultimate victory of good over evil and the end of the present age.*

CHAPTER 11

Pass on Only the Good

THERE IS ABSOLUTELY NO REASON for us to pass on the following to mechanical humans:

- The concepts of hate or malice.
- That the mistreatment and destruction of any entity, certainly including mechanical humans, is OK.

If we're careful and do our job properly, we'll pass on only the concepts and capabilities of being loving beings --- no more evil. Or better yet, mechanical humans will be able to analyze the way we humans treated our fellow man and realize how inhumane, maniacal, irresponsible, senseless, ill-advised, and counterproductive we were. Therefore, they will avoid our abhorrent behavior altogether. In addition, they'll probably also figure out that was the way we were created and we had no control over our odious behavior.

A Voice from the Past Predicts the Future

I mentioned earlier that when I was a kid, I liked to read science fiction. Here's what Isaac Asimov had to say concerning robot behavior in 1942 (from Wikipedia):

Three Laws of Robotics

From Wikipedia, the free encyclopedia

https://en.wikipedia.org/wiki/
Three Laws of Robotics

> **The Three Laws of Robotics** (often shortened to **The Three Laws** or **Three Laws**, also known as **Asimov's Laws** are a set of rules devised by the science fiction author Isaac Asimov. The rules were introduced in his 1942 short story "Runaround," although they had been foreshadowed in a few earlier stories. The Three Laws, quoted as being from the "Handbook of Robotics, 56th Edition, 2058 A.D.," are:
>
> 1. A robot may not injure a human being or, through inaction, allow a human being to come to harm.
> 2. A robot must obey the orders given it by human beings except where such orders would conflict with the First Law.

3. *A robot must protect its own existence as long as such protection does not conflict with the First or Second Laws.*[1]

Now Isaac didn't exactly predict mechanical humans, but notice when the handbook stating the three laws was published:

Handbook of Robotics, 56th Edition, 2058 AD

Isn't it about 2058 AD that true mechanical humans will be fully in charge? It seems we humans have been predicting our evolution and demise for some time now --- without realizing it.

A big smiley face to all of us.

☺ Jack

Jack R Kryder

Dallas Texas

jack@worldnews.earth

worldnews.earth

RESEARCH APPENDIX

Note: If you want to access all the internet URLs listed in this booklet by simply clicking on them, go to my website:

worldnews.earth

==============================

EXTENDED NOTES

This set of extended notes, facts, and comments were collected from the Internet and books. The sources are listed later in the *Research Appendix*. The extended notes are to get you jump-started and give you a peek at what's going on around the world that is responsible for us heading toward extinction. The facts in these notes have been collected, condensed, and put in my own words:

1. This makes it easier and faster for you to get through them.
2. It avoids copyright infringement on my part.

Disappearing Resources

- Humans are using 30 percent more material resources than the Earth can replenish each year. This includes minerals, forests, soils, fresh air, freshwater, arable land, and unpolluted oceans.
- Maybe the most critical to survival is the loss of freshwater.
- Probably the second most critical to survival is the loss of arable land. The world has lost a third of its arable land due to erosion or pollution in the past forty years. And the loss is still occurring due to these natural phenomena. In addition, at the current rates of conversion to other uses, it's estimated we'll lose another 20 percent of our arable land in the next sixteen years. No arable land, no food.
- In addition, the Earth is becoming much less biodiverse. It has been estimated that since 1970, classes such as marine species, freshwater species, tropical forests, dry lands, and grasslands have declined by 33 percent, and populations of

mammals, birds, reptiles, and amphibians have declined by 52 percent.
- By all accounts, if nothing changes in the next twenty years or so, mankind will need several planets to sustain our lifestyle. Basically, we'll be unable to sustain ourselves. This is what gives us the idea/fact that survival of all life on planet Earth is in jeopardy.
- The facts of disappearing resources are a good barometer of what we're doing to our vulnerable Earth. We're ignoring these declines at our own peril.

Disappearing Water

- Oceans cover around 70 percent of the Earth's surface and account for 97 percent of its water.
- Only 3 percent of all water on Earth is freshwater.
- Only about 1 percent of all water found on Earth is easily accessible for human use.
- Water demand already exceeds supply in many parts of the world, and many more areas are expected to experience this imbalance in the near future. In fact, many countries are experiencing moderate to severe water stress on a year-round basis.

Many of these countries are importing half or more of the water they consume.

- 780 million people lack access to clean water and 2.5 _billion_ lack adequate sanitation services.
- Agriculture claims 70 percent of all the freshwater used by humans.
- Some of the causes of water shortages are:
  - Overuse of water tables. Water tables include aquifers and other sources of groundwater. Because the gap between supply and demand is routinely bridged with nonrenewable groundwater, supplies in some major aquifers will be depleted in a matter of decades.
  - Another source of water shortage is pollution of freshwater by cities, agriculture, and industry.
  - A huge amount of water is used in manufacturing processes.
  - _Global warming_ (as in worldwide) is having a huge effect on water shortages. From a weather standpoint, we're getting water and flooding where we can't use it and a lack of water where we need it. Another way of stating it is current areas of precipitation, snowmelt, and streamflow are no longer enough

to supply the multiple competing demands for the world's water needs.

Lack of Food

There has always been a lack of food for a large percentage of the world's population since time began and for many reasons. But it's going to get a lot worse for all of us.

As you know, the main things needed to grow food are arable land and freshwater. Because most arable land has been used to grow food since time began, it is devoid of nutrients. Therefore, it's basically impossible to grow enough food without the use of fertilizers. Fertilizers replace the chemical components that are taken from the soil by growing plants. It requires many types of raw materials to manufacture fertilizers. Fertilizers suck up a large percentage of Earth's resources. The raw materials are supplied to fertilizer manufacturers in bulk quantities of _thousands of tons_. It's impossible to get fertilizers to all areas of the world that need them.

Lack of Food and Water

We all need food and water to survive. Both are becoming scarce at an alarming rate.

Population Problems

Since 1970, the Earth's population has nearly doubled. And it continues to increase at a prodigious rate. No slowdown in sight. Today three-quarters of the world's population live in countries that consume more than they can replenish. We really are in trouble and heading for disaster.

The following websites tie in directly with extinction and evolution and will give you a head start in understanding the reason for this booklet:

Nature Bats Last
Our days are numbered. Passionately pursue a life of excellence.

Resistance is the Only Ethical Response to Near-Term Extinction
May 16, 2013

http://guymcpherson.com/2013/05/
resistance-is-the-only-ethical-
response-to-near-term-extinction/
comment-page-2/#comments

The above website contains many pages and thousands of words. Many are about the coming extinction and what they think we should do about it (see above title). What was very important to me were the brief comments by hundreds of people recognizing, agreeing with extinction facts, and concerned with what's happening. *We are not alone!*

=================================

Voluntary Human Extinction Movement

From Wikipedia, the free encyclopedia

https://en.wikipedia.org/wiki/ Voluntary Human Extinction Movement

The **Voluntary Human Extinction Movement (VHEMT)** is an environmental movement that calls for all people to abstain from reproduction to cause the gradual voluntary extinction of humankind.

VHEMT was founded in 1991 by Les U. Knight, an American activist who became involved in the environmental movement in the 1970s and thereafter concluded that human extinction was the best solution to the problems facing the Earth's biosphere and humanity.

Knight argues that the human population is far greater than the Earth can handle, and that the best thing for Earth's biosphere is for humans to voluntarily cease reproducing. He says that humans are "incompatible with the biosphere and that human existence is causing environmental damage which will eventually bring about the extinction of humans (as well as other organisms)."

VHEMT supports human extinction primarily because, in the group's view, it would prevent environmental degradation. The group states that a decrease in the human population would prevent a significant amount of human-caused suffering. The extinctions of non-human species and the scarcity of resources required by humans are frequently cited by the group as evidence of the harm caused by human overpopulation.

=================================

Terrifying Technology Will Blow Your Mind

https://www.youtube.com/watch?v=JbQeABIoO6A

The above website is a fairly long video with several experts being interviewed. Summary:

- Every important piece of data in the world, of all types, is now on computer systems and stored in what amounts to infinite sized databases.

- Every country that has the ability is cyber-hacking into any and all databases they can detect (which is most databases) --- successfully --- to find out what others know and to use that information to their benefit. This definitely includes governments, companies, and goes all the way down to single individuals.
- There are now very sophisticated drones. A drone has been developed with a basketball-sized camera that has 358 cameras mounted on its surface. It sees all. From 14,000 feet (yes, fourteen thousand feet) it can detect an object as small as six inches and record and recognize faces. They are going to record anything below them. They will be more effective and useful than Google street pictures and what satellites now give us. It has _terabytes_ worth of storage to collect and save all the data. These will eventually be deployed around the world.

What this means to me and this booklet is all this data collection will tell everyone involved where all the minerals are, where all the arable land and food is, and where every drop of freshwater is. We're running out of all four resources at a prodigious rate. When it comes to fighting for survival, this information will be tantamount to life itself.

====================================

How are humans going to become extinct?

By Sean Coughlan BBC News education correspondent

http://www.bbc.com/news/ business-22002530

Last year there were more academic papers published on snowboarding than human extinction.

> *An international team of scientists, mathematicians, and philosophers at Oxford University's Future of Humanity Institute is investigating the biggest dangers. The Swedish-born director of the institute, Nick Bostrom, says the stakes couldn't be higher. If we get it wrong, this could be humanity's final century. Likening it to a dangerous weapon in the hands of a child, he says the advance of technology has overtaken our capacity to control the possible consequences. [Emphasis by author]*

====================================

TOP 10 Amazing Robots in the World 2015 | GOTOP10

https://www.youtube.com/watch?v=YXj0EYlDyxE

There are many organizations/companies working on robots that accurately mimic humans in looks and mannerisms. Then there are the thousands of robots designed to get work done and have no resemblance to humans. Then there are the in-between robots that are humanoid. Visit the above website. They're a hoot. I think you'll get a real kick out of them.

===================================

6 Ways You Know the Robots Are Revolting
[Author --- against humans]

APR 15, 2013 // BY GLENN MCDONALD

http://news.discovery.com/tech/robotics/6-ways-you-know-the-robots-are-revolting-130416.htm

Are we watching the robot revolution take place before our very eyes? Will our future robot overlords look back, 100 years from

now, and regard these moments with digital fondness as they populate their own history books? (Or databases or holo-archives or whatever they have in store?)

SIGNIFICANT BOOKS CONCERNING RESOURCES

===================================

The Land Grabbers: The New Fight over Who Owns the Earth
The Race for What's Left

By Fred Pearce

People and the Planet

by Lyn Sirota

Explores how human activity affects the Earth, e.g., effects of human activity on groundwater and surface water, habitat destruction, construction, emissions, and pollution.

2013

STATE OF THE WORLD
Is Sustainability Still Possible?

"*State of the World 2013* assembles the wisdom and clarity of some of the Earth's finest thinkers, visionaries, and activists into a dazzling array of topics that merge to offer a compellingly lucid and accessible vision of where we are—and what is the wisest and healthiest course for the future." Thirty-four in all.

To access the table of contents, which includes a list of the authors, go to

http://blogs.worldwatch.org/
sustainabilitypossible/state-of-the-world-2013/

THE WONDERFUL, WONDERFUL INTERNET

****

****

WHERE IS ALL THE WATER GOING?

==============================

Freshwater: what's at stake, what we're missing, what we're losing, what it's worth

http://wwf.panda.org/about_our_Earth/about_freshwater/importance_value/

Earth's Disappearing Groundwater

http://Earthobservatory.nasa.gov/blogs/Earthmatters/2014/11/05/Earths-disapearing-groundwater/

Global Warming and Our Shrinking Freshwater Supply

http://www.motherEarthnews.com/nature-and-environment/global-warming-our-shrinking-fresh-water-supply.aspx

Where the world's freshwater is disappearing the fastest

http://knowmore.washingtonpost.com/2015/06/17/where-the-worlds-fresh-water-is-disappearing-the-fastest/

Shoebat Foundation on March 23, 2015
The Nile and Euphrates rivers, and all the countries that rely on them, are in real trouble.

http://shoebat.com/2015/03/23/the-nile-and-the-euphrates-are-drying-up-both-rivers-are-in-the-news-and-both-rivers-are-in-the-bible-an-inevitable-famine-is-plaguing-the-muslim-world/

10 Lakes That Are Disappearing or Already Gone

http://mentalfloss.com/article/56732/10-lakes-are-disappearing-or-already-gone

Will water ever disappear from Earth?

http://sfenvironmentkids.org/water/river_life3.htm

FRESHWATER SYSTEMS

http://www.worldwildlife.org/industries/freshwater-systems

<u>Disappearing **_freshwater_**</u>

https://www.google.com/webhp?sourceid=
chrome-instant&ion=1&espv=2&ie=UTF-
8#q=disappearing%20fresh%20
water

Freshwater Fish are Disappearing: Where is the Global Response?

http://www.livescience.com/49576-freshwater-fish-disappearing.html

As Lake Mead Levels Drop, The West Braces For Bigger Drought Impact

/17/400377057/as-lake-mead-levels-drop-the-west-braces-http://www.npr.org/2015/04for-bigger-drought-impact

Al Gore:

The case for optimism on climate change

http://www.ted.com/talks/al_gore_the_case_for_optimism_on_climate_change?utm_source=newsletter_weekly_2016-02-27&utm_campaign=newsletter_weekly&utm_medium=email&utm_content=talk_of_the_week_image#t-435045

The largest river on Earth is invisible — and airborne

http://ideas.ted.com/this-airborne-river-may-be-the-largest-river-on-Earth/

Off color: 93% of Great Barrier Reef struck by mass coral bleaching event

http://www.gizmag.com/great-barrier-reef-bleaching/42925/?utm_source=Gizmag+Subscribers&utm_campaign=403407a4c6-UA-2235360-4&utm_medium=email&utm_term=0_65b67362bd-403407a4c6-91792181

Is the California drought America's water wake-up call?

http://www.latimes.com/opinion/op-ed/la-oe-famiglietti-chronic-water-scarcity-20160417-story.html

Meathooked and End of Water

http://www.hbo.com/vice/episodes/04/41-meathooked-and-end-of-water/video/ep-405-preview.html?autoplay=true

WHERE IS ALL THE ARABLE LAND GOING?

=================================

Earth has lost a third of arable land in past 40 years, scientists say

http://www.theguardian.com/
environment/2015/dec/02/
arable-land-soil-food-security-shortage

Land Degradation

http://www.globalchange.umich.edu/
globalchange2/current/lectures/land_deg/
land_deg.html

A new vision for agriculture

http://www.momagri.org/UK/agriculture-s-
key-figures/Every-single-day-over-550-acres-
of-agricultural-land-the-equivalent-of-four-
average-size-farms-are-disappearing-in-
France_1057.html

MOST PRODUCTIVE U.S. FARMLAND DISAPPEARING AT FASTEST RATE, REPORT SAYS

https://e360.yale.edu/digest/most-productive-
us-farmland-disappearing-at-fastest-rate-
report-says/2627/

Canada's Disappearing Farmland

http://www.organicagcentre.ca/Newspaper Articles/na_disappearing_farmland_tb.asp

Arable land
From Wikipedia, the free encyclopedia

https://en.wikipedia.org/wiki/Arable_land

Incredible shrinking farmland

http://grist.org/sustainable-farming/2011-11-07-incredible-shrinking-farmland/

World is facing a natural resources crisis worse than financial crunch

http://www.theguardian.com/environment/2008/oct/29/climatechange-endangered habitats

Details about Fertilizer

http://www.madehow.com/Volume-3/Fertilizer.html#ixzz3zs11j0iU

TED Talks: Apocalypse survival guides
How we can make crops survive without water
Open-sourced blueprints for civilization
One seed at a time, protecting the future of food
How to make filthy water drinkable
10 ways the world could end

http://www.ted.com/playlists/334/apocalypse survival guide?utm source=newsletter weekly 2016-02-27&utm campaign=newsletter weekly&utm medium=email&utm content=playlist button

Food Futures Lab

http://www.iftf.org/foodfutures/

Food Futures Projects

http://www.iftf.org/our-work/global-landscape/foodforfuture/

Workable Futures Initiative

http://www.iftf.org/workablefutures/

WHERE ARE ALL THE MINERALS GOING?

===============================

There are at least thirty minerals that are particularly important in the United States for industrial/technological needs. The metals most in demand are: aluminum, copper, gold, silver, iron, tin, platinum, chromium, nickel, lead, and zinc.

Visit the United States Geological Survey (USGS) to find out anything and everything concerning US minerals

http://minerals.usgs.gov/

Rare Earth Minerals' Scarcity Worrisome for Growing Tech Sector

http://www.pbs.org/newshour/bb/business-jan-june10-metals_06-14/

Do We Take Minerals for Granted?

http://minerals.usgs.gov/granted.html#top

List of countries by bauxite production

https://en.wikipedia.org/wiki/List_of_countries_by_bauxite_production

Geology and Nonfuel Mineral Deposits of the United States

http://pubs.usgs.gov/of/2005/1294/a/

Minerals Everywhere and Everyday

http://minerals.usgs.gov/granted.
html#everywhere

Minerals and the Environment

http://minerals.usgs.gov/granted.
html#environment

Future Mineral Supplies

http://minerals.usgs.gov/granted.html#future

Minerals and the Economy

http://minerals.usgs.gov/granted.
html#economy

Mineral Resources Program

The USGS Mineral Resources Program delivers unbiased science and information to understand mineral resource potential, production, consumption, and how minerals interact with the environment.

http://minerals.usgs.gov/index.html

Global Resource Depletion

Is Population the Problem?

http://monthlyreview.org/2013/01/01/global-resource-depletion/

Natural Resources of the Earth

http://www.ecofriendlykids.co.uk/NaturalResourcesEarth.html

Natural Resources - Can We Use Them Forever

http://extension.illinois.edu/world/nres.cfm

Natural resource

https://simple.wikipedia.org/wiki/Natural_resource

Natural Resources

http://www.econlib.org/library/Enc/NaturalResources.html

Characterization and Identification of Critical Mineral Resources

http://minerals.usgs.gov/science/critical-minerals.html

LIVING PLANET REPORT

==================================

Every two years, the Global Footprint Network, WWF International and the Zoological Society of London publish the *Living Planet Report*.

The *Living Planet Report* is the world's leading, science-based analysis on the health of our planet and the impact of human activity. Knowing we only have one planet, the publishers believe that humanity can make better choices that translate into clear benefits for ecology, society and the economy today and in the long term.

The *Living Planet Report* documents the state of the planet—including biodiversity, ecosystems, and demand on natural resources—and what this means for humans and wildlife. The report brings together a variety of research to provide a comprehensive view of the health of the Earth.

The reports are extremely comprehensive and well worth reading. They are available for free in PDF format.

To get the Living Planet Report for 2012, go to:

http://wwf.panda.org/about_our_Earth/all_publications/living_planet_report/2012_lpr/

To get the Living Planet Report for 2014, go to:

http://www.worldwildlife.org/pages/
living-planet-report-2014

Table of Contents

To get past reports, and many other relevant documents, go to:

http://www.footprintnetwork.org/en/index.php/GFN/page/living_planet_report2/

WORLD POPULATION MEDIAN AGES

==

The median age of people around the world is very young. Most people on Earth today are much younger than seventy and thus they will be directly impacted by the problems discussed in this booklet --- especially the lack of freshwater and food.

The World Fact Book
Get the median age of males and females for every country in the world.

https://www.cia.gov/library/publications/the-world-factbook/fields/2177.html

These 8 maps show the median age of every country on Earth

http://theweek.com/articles/443122/8-maps-show-median-age-every-country-earth

WHERE IS ALL THE MONEY GOING?

================================

The National Debt With a "Noble" Solution

http://www.federalbudget.com/noble.html

Can A Nation $19 _Trillion_ In Debt Afford Higher Interest Rates & Will This Change Our Retirements?

http://danielamerman.com/va/Conflict.html

United States Balance of Trade 1950-2016

http://www.tradingeconomics.com/united-states/balance-of-trade

Fighting for a U.S. federal budget that works for all Americans

https://www.nationalpriorities.org/campaigns/us-federal-debt-what/?gclid=CK_QvPzAg8wCFYGFaQodA8wBqw

Federal Spending: Where Does the Money Go

Federal Budget 101

https://www.nationalpriorities.org/budget-basics/federal-budget-101/spending/

5 facts about the national debt: What you should know

http://www.pewresearch.org/fact-tank/2013/10/09/5-facts-about-the-national-debt-what-you-should-know/

United States Total National Debt

http://www.concordcoalition.org/us-total-national-debt?gclid=CIq748bCg8wCFdgSgQodeIQDLQ

U.S. and World Population Clock

http://www.census.gov/popclock/

NEW TECHNOLOGY _IS_ COMING ON STRONG

===============================

How are humans going to become extinct?
By Sean Coughlan BBC News education correspondent
[Aside: By technology taking over.]

http://www.bbc.com/news/business-22002530

Last year there were more academic papers published on snowboarding than human extinction.

> An international team of scientists, mathematicians, and philosophers at Oxford University's Future of Humanity Institute is investigating the biggest dangers. The Swedish-born director of the institute, Nick Bostrom, says the stakes couldn't be higher. If we get it wrong, this could be humanity's final century. Likening it to a dangerous weapon in the hands of a child, he says _the advance of technology has overtaken our capacity to control the possible consequences_. [Emphasis by author]

Next Future Terrifying Technology Will Blow Your Mind

https://www.youtube.com/watch?v=JbQeABIoO6A

The above website is a fairly long video with several experts being interviewed. Summary:

- Every important piece of data in the world, of all types, is now on computer systems and stored in what amounts to infinite sized databases.
- Every _country_ that has the ability is cyber-hacking into any and all databases they can detect (which is most databases) --- successfully --- to find out what others know and to use that information to their benefit. This definitely includes governments, companies, and goes all the way down to single individuals.
- There are now very sophisticated drones. A drone has been developed with a basketball-sized camera that has 358 cameras mounted on its surface. It sees all. From 14,000 feet (yes, fourteen thousand feet) it can detect an object as small as six inches and record and recognize faces. They are going to record anything below them. They will be more effective and useful than Google street pictures and what satellites now give us. It has terabytes worth of storage to collect and save all the data. These will eventually be deployed around the world.

What this means to me and this booklet is all this data collection will tell everyone involved where all the minerals are, where all the arable land and food is, and where every drop of freshwater is. We're running out of all four resources at a prodigious rate. When it comes to fighting for survival, this information will be tantamount to life itself.

Liquid metal runs through new flexible circuits

http://www.gizmag.com/epfl-stretchable-circuits/42115/?utm_source=Gizmag+Sub scribers&utm_campaign=0eb3f72b36-UA-223 5360-4&utm_medium=email&utm_term=0_ 65b67362bd-0eb3f72b36-91792181

Monkeys master thought-controlled wheelchair

http://www.gizmag.com/monkey-wheelchair-brains/42162/?utm_source=Gizmag+Sub scribers&utm_campaign=deb8d2ac2d-UA-223 5360-4&utm_medium=email&utm_term=0_ 65b67362bd-deb8d2ac2d-91792181chrome:// newtab/

Electron state-changing device could be used in quantum computing

http://www.gizmag.com/electron-state-change-device-quantum-computing/41913/?li source=LI&li medium=default-widget

Transistor count

https://en.wikipedia.org/wiki/Transistor count

HTC Vive review: This is VR the way we always imagined it

http://www.gizmag.com/htc-vive-review/42614/?utm source=Gizmag+Sub scribers&utm campaign=1ddc27f753-UA-223 5360-4&utm medium=email&utm term=0 65b67362bd-1ddc27f753-91792181

Oculus Rift review: Polished, often magical ... but ultimately second best

http://www.gizmag.com/oculus-rift-review/42846/

LaCie goes big and roomy with 96 TB hard drive

http://www.gizmag.com/lacie-12-big-thunder bolt-hard-drive/42878/?utm source= Gizmag+Subscribers&utm campaign=

2c4c0e488f-UA-2235360-4&utm_medium=email&utm_term=0_65b67362bd-2c4c0e488f-91792181

Super flat material could extend life of Moore's Law

http://www.gizmag.com/2d-semiconductor-tin-monoxide/41843/?utm_source=Gizmag+Subscribers&utm_campaign=ef6fe510de-UA-2235360-4&utm_medium=email&utm_term=0_65b67362bd-ef6fe510de-91792181

Nanotubes Serve as Light Emitter in Integrated Photonic Circuit

http://spectrum.ieee.org/nanoclast/semiconductors/optoelectronics/nanotubes-serve-as-light-emitter-in-integrated-photonic-circuit

Researchers Use Coal to Make Electronics

http://www.laboratoryequipment.com/news/2016/04/researchers-use-coal-make-electronics

Mapping the thesaurus of the human brain

http://www.gizmag.com/brain-thesaurus-language-map/43056/?utm_source=Gizmag+Subscribers&utm_campaign=e2a9b37370-UA-2235360-4&utm_medium=email&utm_term=0_65b67362bd-e2a9b37370-91792181

Chance discovery puts graphene electronics closer to mass production

http://www.gizmag.com/graphene-doping-mass-production/41829/?utm_source=Gizmag+Subscribers&utm_campaign=ef6fe510de-UA-2235360-4&utm_medium=email&utm_term=0_65b67362bd-ef6fe510de-91792181

Eternal data storage demonstrated in nanostructured glass

http://www.gizmag.com/eternal-data-storage-nanostructured-glass/41951/

5D GLASS DISC CAN STORE 360TB (terabytes)!

http://hackaday.com/2016/02/18/5d-glass-disc-can-store-360tb/

Eternal 5D data storage could record the history of humankind

http://phys.org/news/2016-02-eternal-5d-storage-history-humankind.html

NVIDIA Tesla P100 The Most Advanced Datacenter GPU Ever Built

http://www.nvidia.com/object/tesla-p100.html

Intel's 10-Core Monster Planned For Q2 2016 Debut, According to New Leak

http://www.digitaltrends.com/computing/broadwell-e-roadmap-leak/

A history of Intel microprocessor (CPU) development

https://en.wikipedia.org/wiki/
List of Intel microprocessors

Top 10 Most Influential Tech Advances Of The Decade

http://www.pcmag.com/article2/0,2817,2374825,00.asp

Coal-based electronics: A potential usurper to silicon's throne?

http://www.gizmag.com/nanomaterial-coal-carbon-electronics-mit/42905/?utm_source=Gizmag+Subscribers&utm_campaign=403407a4c6-UA-2235360-4&utm_medium=email&utm_term=0_65b67362bd-403407a4c6-91792181

MECHANICAL HUMANS _ARE_ COMING

=============================

Developmental robotics
Great Wikipedia article.

https://en.wikipedia.org/wiki/
Developmental_robotics

If consciousness is an algorithm, then a robot can be conscious

(Ethical dilemmas may arrive sooner than we think)

http://www.theguardian.com/theobserver/
2016/mar/20/
observer-letters-artificial-intelligence

Will humanoid robots build tomorrow's aircraft?

http://www.gizmag.com/humanoid-robots-
aircraft-assembly/41846/?utm_source=
Gizmag+Subscribers&utm_campaign=
718f306345-UA-2235360-4&utm_medium=
email&utm_term=0_65b67362bd-718f30
6345-91792181

TOP 10 Amazing Robots in the World 2015 | GOTOP10

https://www.youtube.com/watch?v=YXj0EYlDyxE

There are many organizations/companies working on robots that accurately mimic humans in looks and mannerisms. Then there are the thousands of robots designed to get work done and have no resemblance to humans. Then there are the in-between robots that are humanoid. Visit the above website. They're a hoot. I think you'll get a real kick out of them.

6 Ways You Know the Robots Are Revolting
[Author --- against humans]

APR 15, 2013 // BY GLENN MCDONALD

http://news.discovery.com/tech/robotics/6-ways-you-know-the-robots-are-revolting-130416.htm

*Every few weeks, it seems, a **new story** hits the headlines about the latest breakthrough in the field of robotics. Innovations are happening all the time, of course. The term "robotics" encompasses a global system of technology, industry*

and research. But certain kinds of robot stories break through into popular culture with alarming regularity.

Taken together, these stories can provoke a certain unease. Read enough of them in a row, and the question presents itself: Are we watching the robot revolution take place before our very eyes? Will our future robot overlords look back, 100 years from now, and regard these moments with digital fondness as they populate their own history books? (Or databases or holo-archives or whatever they have in store?)

Forthwith, a look at recent developments in robotics that could well prove historical, albeit for all the wrong reasons.

Nadine the eerie social robot looks and feels like humans do

http://www.gizmag.com/nadine-social-robot-humans/41089/?utm_source=Gizmag+Subscribers&utm_campaign=61511319e2-UA-2235360-4&utm_medium=email&utm_term=0_65b67362bd-61511319e2-91792181

NASA Developing Robots with Human Traits

http://www.nasa.gov/vision/universe/roboticexplorers/robots_human_coop.html

Cutting Edge
Developing the robot of the future

http://www.latimes.com/business/technology/
la-fi-cutting-edge-ucsd-robots-20151101-
story.html

Developing robots for the hospital emergency
room

http://news.vanderbilt.edu/2010/12/
developing-robots-for-the-hospital-
emergency-room/

Develop Robots

https://www.facebook.com/developrobots

ESA Developing Robots **for Space Exploration**

http://www.cadincadout.com/
esa-developing-robots-for-space-exploration/

Endeavor Robotics – dedicated to serving
customers in the defense, public safety, energy,
and industrial markets

http://endeavorrobotics.com/

THE COMPLEX PROCESS OF DEVELOPING INTELLIGENCE IN ROBOTS

http://beckman.illinois.edu/news/2015/05/intelligence-in-robots

ROBOTIC DEVELOPMENT

https://www.robots.com/articles/viewing/robotic-development

VGo. From anywhere. Go anywhere.
VGo replicates a person in a distant location.

http://www.vgocom.com/

Robots

http://www.livescience.com/topics/robots/

Find out everything there is to know about robots and stay updated on the latest robots and inventions with the comprehensive articles and interactive features. Learn more about man's innovative creations as people and scientists continue to invent and interact with robots.

ROBOTICS

Slithering serpentine robot snakes its way to seabed inspections

http://www.gizmag.com/snake-robot-eelume-seabed-inspections/42874/

News about Robots

https://www.theguardian.com/technology/robots

One man's opinion: Rise of the robots will harm the Earth as well as humans

http://www.theguardian.com/theobserver/2016/mar/27/leters-robots-energy-consuming

Is it a robot? Is it a phone? Yes it's both! Introducing RoBoHoN

https://www.theguardian.com/technology/video/2016/apr/14/is-it-a-robot-is-it-a-phone-yes-its-both-introducing-robohon-video

10 Amazing Robots That Will Change the World

https://www.youtube.com/watch?v=6feEE716UEk

The Most Awesome Robots

Will robots take over the world one day in your opinion?

https://www.youtube.com/
watch?v=S5AnWzjHtWA

Developing robots that are self-aware

http://www.cnn.com/videos/tv/2015/08/26/
spc-make-create-innovate-robots-self-
aware.cnn

Samsung developing robots to replace cheap Chinese labor

http://www.wired.co.uk/
news/archive/2015-10/19/
samsung-south-korea-robots-cheap-labour

Simplified system could allow for better robot-human communications

http://www.gizmag.com/mit-streamlined-
robot-human-team-communications/41879/
?utm_source=Gizmag+Subscribers&utm_
campaign=3824de6a6f-UA-2235360-4&u
tm_medium=email&utm_term=0_
65b67362bd-3824de6a6f-91792181

IMPROVING THE EARTH

==============================

January 2013
Excerpt from:
Global Resource Depletion
by Fred Magdoff

http://monthlyreview.org/2013/01/01/
global-resource-depletion/

This is a 6,800-word article that covers:

- Is Population the Problem?
- Resource Depletion and Overuse
- The Accumulation of Capital is the Accumulation of Environmental Degradation
- Let's Talk Population
- Chart 1. Share of World Consumption by Income Decile
- Population Declines and Capitalist Economies
- Combating Pollution and Resource Depletion/Misuse
- Thirty-one supporting notes

This article is very well written and is far more detailed and wide-ranging than the headings indicate. If you're in any way interested in global resource depletion, then you owe it to yourself to read the entire article.

Excerpt:

As a result, it is important to clarify a number of such issues and get potential stumbling blocks, related to population specifically, out of the way before continuing with this part of the discussion. Our starting points should be:

- *All people everywhere should have easy access to medical care, including contraceptive and other reproductive assistance.*
- *As living standards rise to a level that supplies family security, the number of children per family tends to decline. But, depending on the circumstances, there may be good reasons for poor women and men to have fewer children even before they have more secure futures and for individual countries to encourage smaller families.*
- *There <u>are</u> poor countries where overgrazing, excess logging of forests, and soil degradation on marginal agricultural land are caused by relatively large populations and the lack of alternate ways for people to make a living except from*

the land. *This problem may be worsened by the low yields commonly obtained from infertile tropical soils. But we also need to recognize that these problems are not only an issue of population density.*

- *Some countries have populations so large relative to their agricultural land that importing of food will be needed into the foreseeable future. One of the largest of these nations is Egypt, with a population of over 80 million people and arable land of 0.04 hectares (less than one-tenth of an acre) per capita. These countries are condemned to suffer the consequences of rapid international market price hikes that occur frequently and of having to maintain significant exports just to be able to get sufficient hard currency to import food. There are other countries—such as Saudi Arabia, the United Arab Emirates, Oman, and Qatar—that have a larger population than what can be sustained by available water/food resources, but each of them can currently use oil and/or other commercial income to obtain sufficient food for their populations.*

- *All else being equal—which, of course, it never is—larger populations on the Earth create more <u>potential</u> environmental problems. So population is always an* environmental factor—though usually not the main one, given that economic growth generally outweighs population growth and environmental degradation arises mainly from the rich rather than the poor.

- If we assume that all people will live at a particular standard of living, there is a finite carrying capacity of the Earth, above which population growth will not be sustainable because of rapid depletion of too many resources and too much pollution. For example, it is impossible for all those currently alive to live at what is called a "Western middle-class standard"— for to do so we would need more than four Earths to supply the resources and assimilate pollutants.
- There are currently approximately 7 billion people in the world and, given current trends, the population is expected to be around 9 billion in 2050, and over 10 billion by 2100.

Printed in the United States
By Bookmasters